Boris Kastel

Neptune's spear

Éditions Muse

Publisher:
Éditions Muse
is a trademark of
International Book Market Service Ltd., member of OmniScriptum Publishing Group
17 Meldrum Street, Beau Bassin 71504, Mauritius

Printed at: see last page
ISBN: 978-620-2-29336-5

Modern Montenegrin Poetry

Boris Jovanović Kastel

NEPTUNE'S SPEAR

Translated and edited

by Vladimir Sekulic

Boris Jovanović Kastel

CONFIDENCE

I don't trust the sea anymore

it did not withdraw before us

to the wine bottle of the antique shop

or the aquarium of Peter the second Orseol,

nor has it without reasoning flooded us,

glittering and murmuring

it plays kolo* without a leader,

to a hundred year old circle

and bacchanals with a Lovćen fary

it lights.

* Montenegrin folk dance

THE BANNED

She rushed to the sun long ago

and it celebrates or burns down.

They make me forget her,

but I can't

because the sun is still rising

above the Mother of Jesus in Perast

where in the cell

surrounded by the senses of panihidas

by the storms and turnkeys

I hear the burning of the eagle

at the carnival of merchants.

I survive by biting my nails

and I secretly drink diluted urine,

by the fish skeleton

I engrave the genealogy

of gentlemen and haiduks

of cut veins.

Excuse me the lady of Montenegro,

I read and remember you –

banned to the promise of sandy covers.

A GIRL FROM NAPLES

I followed her for a whole hour

along the streets of Naples.

I remember the market place,

the portals of the old town,

the strands without strollers.

I followed her so barefoot

with a transparent skirt and a blouse,

without a brassier

As if she sensed me,

she turned toward me,

with woman's shrewdness,

she looked at the tower.

I thought to stop her,

to introduce myself,

to tell her that I am a poet

from a country Montenegro

and without hesitation

to declare her love.

Or maybe

to give her a book,

to offer her a stroll

to the cathedral

and to wake up in the attic

of a rented apartment.

I was silent,

and lost courage

and I looked at the tower

at a minute to noon.

I didn't ask her anything

and I didn't even think of touching her

not even her shadow –

as a half of a violin –

played by a blind man

on the other side of the street.

She was leaving

as if she knew

who I was,

from which country –

where people consciously

hold their eyes closed

and blind people

feel the presentiment

of the end.

PYROTECHNICS

A poem instead of a pill for insomnia, is not for girls to get married, puritans and anthologists with a jacket instead of a pleura

Pyrotechnics of the sea senses

illuminates the warrior without an altar,

a hand of procedures and shield,

the last defense against the recent outgrow.

With the bride of a bordello, a traveling

companion

with a bust illuminated by pyrotechnics

of stricken waves,

the other side of a being

horror struck by laughter.

Salvation, the key for the libido of gates,

running away from the high tides of

pragmatism,

the trade with oneself:

to steal her bosom,

her nipple is a stamp

of a charter of a permit

for the fortified city

of the southern south.

Between the body and the most holy hearth of Yesus 2000. By the palace of the family Paskvali in the hysteria of the raining of Kotor heaven.

CHESS

A canon opposite a king

is withdrawing

before the slandered Lord.

Such canons are rare,

there are no museums, human ones.

It is too much to speak

about hunters for heads,

about flushed horses,

about the clowns pawns

who are secretly waiting

for the evening gown of the queen.

GENERATIONS

For the minority for ever

Jimenez

Here the sailors burn the ships,

saving the ship's logs

written with colorless years,

trusting the harbor fortune-tellers,

not feeling the accident in the storm...

Here the women pull in the bosom

and in short skirts begin to move

to the portals

to satisfy the emigrant workers and tramps.

Academicians, actors, poets

and other wax figures

dream about other countries –

under the splendor of the lampions

they hurry to go

that their talents don't turn into stones.

They go

knowing that proud and damnation

of sculptures they can't have.

Generations

starched and free on the road backward

stand by the sea,

acting and toasting in black robes

to future, looking at the towers

whose oldness despises their memories.

The oranges are drying, thunders are warning

the rocks which made me drunk

and cut me out

and don't seduce the heartless.

I am watching the fight of the doves

for a crumb from the pavement,

they are conscious of the goal,

survival…

I don't trust a sad person

while he is a bird

and those boys who brought

the ships of paper to the open sea

who suspect the tide of reason

sending the paper

where the exiled will register

the mistakes of the majority.

THE CLOCKS

The wings of schizophrenic gulls

a minute to twelve

don't show up any more.

Behind unknown, blind ships,

without the eighth passengers,

charmed by going out,

they went to no return.

The sea hot tempered in the chest.

On tops of towers and castles

decorated by God weaving the flags

at half masts, a minute to noon.

The new illusion will not come –

for a long time,

nor it will hurry when the high sea needs

the blood group of Citera, Cicero's lady

after a meal and a prostitute from the highest

society –for salvation.

Rusty of tears of angels

for a revived and again petrified caryatid,

the mechanisms of clocks have remained.

Rusty - for us to follow them.

A PALM TREE

In the night of the first day of summer

from a museum of the southern museums

the Neptune's spear was stolen,

a young palm tree was broken

from a tree lined path

without an end.

It is the second night of summer,

I am the witness –

a spineless and hunchback person.

THE SECOND DAY

The turbulent sea

follows the thought of the fatherland,

I remember the ostensible rebels –

It was the calm sea,

going for figs in the nearby garden

I was thinking of exile

the medicine for dignity

for myself, before the others…

the sea answers by tide,

I offer fruit to my girlfriend, I bite her ear,

we read the Petrarca's sonnets,

her bosom, older fig relatives

crossed legs, a plea for a walk

make me think of fidelity

in infinite depths disappeared,

long ago and long ago…

SEAGULLS

He was looking for his shadow
which he had lost in the distance.
As he was reading the parchments of the
inventor
Ctesibius on his water clock,
walking barefoot on the shore without seabirds
and leaving after himself a shadow of seagulls
with folded wings.
Fishermen, priests and Ulysses' bastards
in blue T-shirts with a bal l in their hands
greeted him—Good morning, master.
He answered with a cry of a seagull
scared both by himself and the croaking.
For centuries he lazily
looked for his shadow
but in fact, being a martyr, he never lost it.

THE MORNING WHEN THE HORIZON CHANGED HIS PERSONALITY

He was leaning out of a barque.
He took a deep look at the surface stretching to
the horizon
more peaceful than the wistfulness of the South
and clearer than Venetian glass
to count his patches of grey hair and wrinkles.
But
in the image on the surface he sighted the soul
of a middle-aged poet, a conspirator
at Caesar Octavianus's court.
He took fright and squinted his eyes.
Believing that he had got drunk
on the scent of lemons at the port
or on the liquor from carob pods,
he rowed back
to the threshold of the palace.
In his bed the cat was waiting for him.

Yet,

as he was lying down to sleep,

a laurel wreath fell off his head.

There was not a sole mirror

in the whole world.

Ever since that night he could only look at

himself

on Roman coins.

Translated by Petr Anténe

Czech-English anthology of contemporary Slavic poetry *Bludné kořeny / Wayward Roots* that was published in 2014 at Palacký University, Olomouc, Czech Republic.

BIOGRAPHY

Boris Jovanovic Kastel (Trebinje, 1971), considered by literary critics the most important Montenegrin poet of Mediterranean origin and prominent name of Mediterranen poetry.

He published the following books of poetry:

The Scents of regrets (1994)

The Rings of Seaside (1995)

Footnotes of Southern Bells (1997)

The Anatomy of The Mediterranean Day (1998)

The Mediterranean Agenda and Predicting the past (2000)

The Mediterranean Hexateuch (2003)

The Ego of the Sea (2004)

Wedding with Cuttle-fish (2007)

Neptune's spear (Selected poems in English, Podgorica, Montenegro, 2007)

Mediterranean indigo (2008, selected poems)

Lunch on the cliff (Podgorica, Montenegro, 2010)

Kosilo na čeri (2014, selection from poetry in Slovenian language)

A man without mainland (Podgorica, Montenegro, 2015)

A letter of invitation to the sun (Podgorica, Montenegro, 2016)

Sea in the hands (Novi Sad, Serbia, 2017)

Zeus in Budva Casino (Budva, Montenegro, 2017)

Are we waiting for boats (Selection from poetry in Macedonian language, Bitola, Macedonia, 2018)

Reefs from bones (Beograd, Serbia, 2018)

Testament in mussels (Opatija, Croatia,2019)

He also published four books of selected essays:

The Parchement of Mermaid's bust (2000)

The Fifth Side of South (2005)

Mirroring the Calm (2009)

The Mediterranean Enlightenment – Our

Mediterranean, Compass of Faith (2012)

Mediterranean Nobleman, the book with selected essays of local and foreign writers on poetry of Boris Jovanovic Kastel (2010).

He won the Nosside World Poetry Praise, awarded under the auspices of UNESCO's World Poetry Directorate in Reggio di Calabria (2011).

He was editor of the Montenegrin literature review Ovdje (2000-2003).

He published his essays in the Montenegrin daily news Pobjeda, for many years.

His poetry was translated in Italian, English, Polish, Czech, Hungarian, Albanian and Slovenian.

His poetry is presented in Antology of the World Poetry Nosside, in Italian, as well as in Antology of Mediterranean Love Poetry from the oldest time until nowdays, Antology of the Slavic poetry in Slovenian, Antology of the Montenegrin poetry in Italian, Antology of the Montenegrin poetry in Albanian, several antologies of Montenegrin poetry on wine, women…

A book of poetry in Slovene Lunch on the cliff has announced the 2014 Slovenian publishing house Hiša poezije anthology edition of European poets Poetikonove lire as part of the European Commission. Book of poetry Letter of invitation to the sun has been published in the prestigious edition of the Montenegrin poetry Savremenik the Institute for textbooks and teaching aids from Podgorica.

Poetry prizes – Literary feather for the best book of the Year (Croatian Literary Society, Rijeka, Croatia, 2016), Kniževni branovi (Literary encounters in Struga, Macedonia, 2017), Balkan Jewelry (Institute for Humanities Belko, Belgrade, Serbia) and prize Goran Bujić (Croatian Literary Society, Zadar, Croatia, 2018) for contribution to the poetry of the Mediterranean.

Italian magazine of world literature and other arts Margutte, in the October issue of 2014, he published poetry Boris Jovanovic Kastel in translation Maria Teresa Albano.

Selected as one of thirty poets of the world for The VI World Poetry Festival in Calcuta (India, 2012).

He lives in Podgorica.

Afterword

B. J. Kastel has, unlike his piers, immediately joined the group of distinguished poets. Indeed, many great poets had their greatest successes precisely when they were young. Boris Jovanović Kastel immediately entered the world of literature in a grand manner. Soon, he became one of our most significant poets. It is already well known that his source of inspiration is the Mediterranean, which can give a lot in every way, especially in spiritual sense. Boris Jovanović Kastel has grown into a free-minded individual and a great author. He became all that thanks to his reading, knowledge and great talent, but also thanks to the Mediterranean, as the area where he exists and travels through time and space, usually alone, avoiding any flock.

Vladimir Sekulic

THE SEA AS A MEANS OF POETIC EXISTENCE

The poetry of Boris Jovanovic Kastel

The sea as a space of mythological timelessness, the sea as a means of poetic existence, maritime culture as a source of timeless inspiration – such are the Mediterranean poetics of the Mediterranean poet Boris Jovanović Kastel. The poem "Seagulls" mentions centuries of searching for a shadow – he who aspires for longevity should enter the world of poetry. In the poem he talks about the "parchments of the inventor Ctesibius on water clocks" – and then enter "fishermen, priests, and Odysseus's bastards / in blue T-shirts with a ball in their hands" – this is the world of Mediterranean ports – ancient history shakes wings with the modern presence. And the seagulls rub their hands in glee. Kastel is mining the rich history of Montenegro, for its capital city of Podgorica (formerly Titograd) is built upon the ancient city of Doclea (Duklja); the Roman Emperor Diocletian was from the area. There are traces of many civilisations here: ancient Greece, Rome, Byzantium, the Venetian Republic, and the Ottoman Empire. Today's Montenegro is still quite multicultural. Kastel demonstrates his timelessness inconspicuously, and for us here invisibly on the linguistic level, when in his verses he does not use the modern Montenegrin symbols of ś and ź, but writes using the more universal forms of sj and zj. Perhaps because his Montenegrin identity is anchored deep in the Mediterranean cultural legacy, his language keeps its historic patina; perhaps it is an expression of defiance against the passing of language laws – after all, a poet of the imperial court insures the correctness of his words with his own throat.

Vladimir Sekulić notes that Kastel's "source of inspiration is the Mediterranean, which can give a lot in every way, especially in spiritual sense. (…) [T]he Mediterranean, [is] the area where he exists and travels through time and space, usually alone, avoiding any flock." The special loneliness and typical Kastelian dislocation and relocation in time echoes through the poem "The Morning When the Horizon Changed His Personality" – the historicising images of a barque, Venetian glass, a laurel wreath (unexpectedly falling from his head) and Roman coins are backdrops for a scene of inner transformation: "in the image on the surface he sighted the soul / of a middle-aged poet, a conspirator / at Caesar Octavianus's court." The lyrical subject is inexorably thrown into history.

In the world there are no mirrors – it is only possible to gaze at the surface, however that "inexorably" reveals the interior, which is not tied to the passage of time. We don't age in grey hair and wrinkles on the future axis, we age spiritually when we are stuck in the past; and yet we can dip past the surface reflection into pre-history: myth, the history which has not been a part of our own physical being, can be a fountain of refreshing youth to draw upon. These are the mythological depths of Kastel's Mediterranean. Those who would like to dip into more of Kastel's poetry, these little poetic time machines, will appreciate the translations of his work into other Slavic languages and into English on his website.

The two poems here will be published in 2014 in a col lection entitled Beskopnik, or "The Man Who Has Never Seen the Mainland" – a one-word concept hard for landlocked Czechs to get their heads around. (Here I would like to also mention the literary signpost Potápěč [Diver], the "breathing apparatus for world literature" by the Czech poet and editor Pavel Kotrle.

On his pages you can find digital links not only to Montenegrin literature, but also to some 180 national literatures.)

Robert Hýsek & Matthew Sweney

Czech-English anthology of contemporary Slavic poetry *Bludné kořeny* / Wayward Roots that was published in 2014 at Palacký University, Olomouc, Czech Republic.

Table of Contents

Druck:
Canon Deutschland Business Services GmbH
im Auftrag der KNV-Gruppe
Ferdinand-Jühlke-Str. 7
99095 Erfurt